UPDOG

MILITARY MACHINES
HELICOPTERS
in Action

MARI BOLTE

Lerner Publications ◆ Minneapolis

Copyright © 2024 by Lerner Publishing Group, Inc.

All rights reserved. International copyright secured. No part of this book may be reproduced, stored in a retrieval system, or transmitted in any form or by any means—electronic, mechanical, photocopying, recording, or otherwise—without the prior written permission of Lerner Publishing Group, Inc., except for the inclusion of brief quotations in an acknowledged review.

Lerner Publications Company
An imprint of Lerner Publishing Group, Inc.
241 First Avenue North
Minneapolis, MN 55401 USA

For reading levels and more information, look up this title at www.lernerbooks.com.

Main body text set in Aptifer Sans LT Pro.
Typeface provided by Adobe Systems.

Library of Congress Cataloging-in-Publication Data

Names: Bolte, Mari, author.
Title: Helicopters in action / Mari Bolte.
Description: Minneapolis : Lerner Publications, [2024] | Series: Military machines (Updog books) | Includes bibliographical references and index. | Audience: Ages 8–11 | Audience: Grades 4–6 | Summary: "Military helicopters are useful in combat, but they are just as important away from the battlefield. Learn about how helicopters provide disaster relief and assist in search and rescue missions"— Provided by publisher.
Identifiers: LCCN 2022043398 (print) | LCCN 2022043399 (ebook) | ISBN 9781728491707 (library binding) | ISBN 9798765603406 (paperback) | ISBN 9781728498942 (ebook)
Subjects: LCSH: Military helicopters—United States—Juvenile literature.
Classification: LCC UG1233 .B658 2024 (print) | LCC UG1233 (ebook) | DDC 358.4/1830973—dc23/eng/20230104

LC record available at https://lccn.loc.gov/2022043398
LC ebook record available at https://lccn.loc.gov/2022043399

Manufactured in the United States of America
1 – CG – 7/15/23

Table of Contents

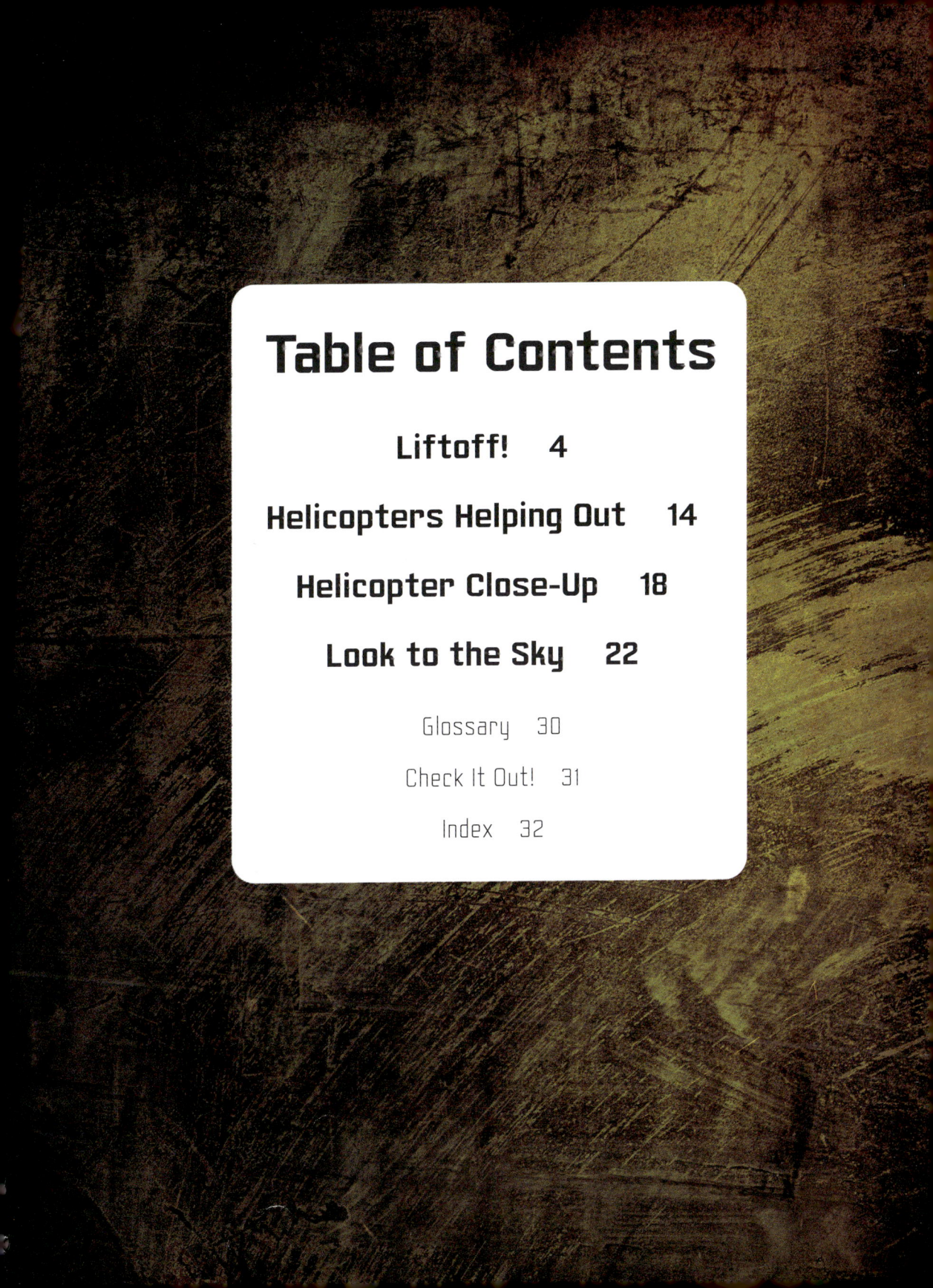

Liftoff!   4

Helicopters Helping Out   14

Helicopter Close-Up   18

Look to the Sky   22

Glossary   30
Check It Out!   31
Index   32

LIFTOFF!

A helicopter flies overhead. It swoops close to the ground, looking for a safe place to land. Finally, the pilot lowers the helicopter to the ground.

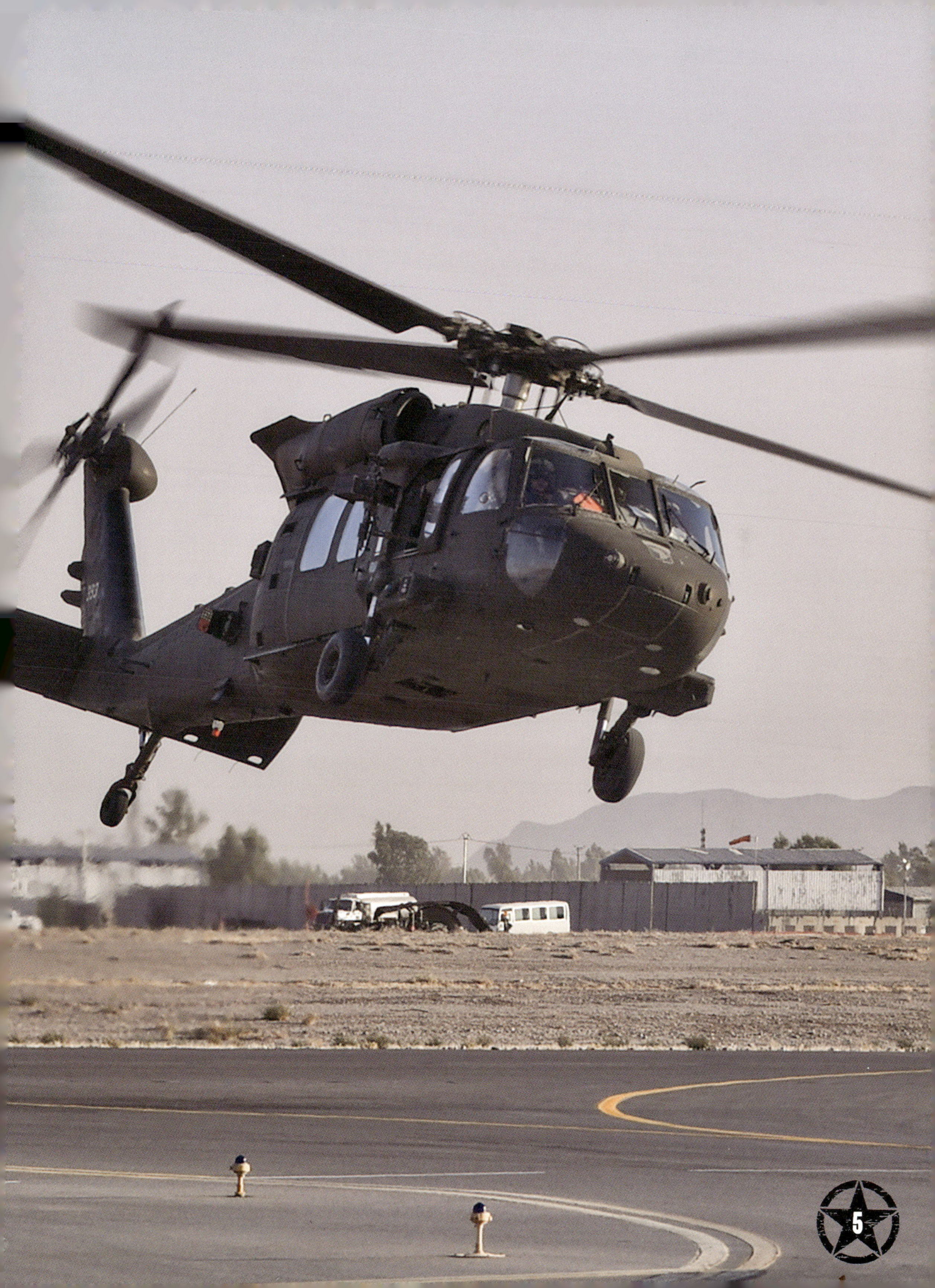

Soldiers quickly jump out of the helicopter. They unload supplies.

Then they run off to begin their mission.

The pilot guides the helicopter back into the air.

It will wait nearby until it is needed again.

Larger helicopters have two pilots. One flies the vehicle while the copilot helps navigate.

The copilot also uses the helicopter's weapons in combat.

The fastest helicopters can fly at speeds of almost 300 miles (483 km) per hour.

The largest helicopters can carry 44,000 pounds (19,958 kg) of supplies.

UP NEXT! LOWER THE LANDING GEAR.

HELICOPTERS HELPING OUT

Five branches of the US military use helicopters. Some attack enemies. Others carry people or supplies where they need to go.

15

Military helicopters often help during natural disasters, such as forest fires.

The helicopters carry huge buckets of water through the air. As they fly over the forest, they dump the water on the flames.

HELICOPTER
Close-Up

Tail Rotor

Landing Skids

Main Rotor

Cockpit

Military helicopters can also help find people who are lost. They can fly low over a forest or the ocean.

Helicopters can safely fly into places where people on foot cannot go.

UP NEXT! BUCKLE UP.

21

Look to the Sky

The military often names helicopters after Native American tribes or people. It is a tradition that began in the 1940s.

23

Apache helicopters are used in combat.

More than 1,200 Apaches have flown over four million mission hours since 1984.

With two main rotors, Chinooks look different than an ordinary helicopter.

They can carry soldiers, weapons, equipment, and fuel. Chinooks are also used to help during natural disasters.

The Sea King is used to transport important people, such as the president of the United States.

Helicopters stand ready to take off and serve whenever they are called for duty.

29

Glossary

branch: a part of the military

combat: fighting

copilot: a person who assists in flying an airplane or helicopter

navigate: to steer a course in a ship or aircraft

rotor: the rotating part of a helicopter that helps it achieve liftoff

Check It Out!

Austin, Mike. *Hooray for Helpers! First Responders and More Heroes in Action*. New York: Random House, 2020.

Britannica Kids: Helicopter
https://kids.britannica.com/kids/article/helicopter/390246

Cella, Clara. *Fighter Pilots*. Minneapolis: Lerner Publications, 2023.

DK Find Out: Helicopters
https://www.dkfindout.com/us/transportation/history-aircraft/helicopters/

KonnectHQ: Facts About Helicopters
https://www.konnecthq.com/helicopters/

Leed, Percy. *The US Marine Corps in Action*. Minneapolis: Lerner Publications, 2023.

Index

Apaches, 24–25

Chinooks, 26–27

natural disasters, 16, 27

pilots, 4, 8, 10–11

soldiers, 6, 27

weapons, 11, 27

Photo Acknowledgments

Image credits: guvendemir/Getty Images, p.4; Sgt. Steven Lopez/DVIDS, p.5; Lance Cpl. Kyle Chan/DVIDS, p.6; Spc. Avery Cunningham/DVIDS, p.7; guvendemir/Getty Images, p.8; guvendemir/Getty Images, p.9; Capt. Brian Harris/DVIDS, p.10; Chief Warrant Officer Cameron Roxberry/DVIDS, p.11; Robert Sullivan/flickr, p.12; Lance Cpl. Elias Pimentel/DVIDS, p.13; lloyd-horgan/Getty Images, p.14; kojihirano/Shutterstock, p.15; Tech. Sgt. Christopher Milbrodt/DVIDS, p.16; Stocktrek Images/Getty Images, p.17; grahamheywood/Getty Images, p.18; Stocktrek/Getty Images, p.20; sierrarat/Getty Images, p.21; Ofer Zidon/Stocktrek Images/Getty Images, p.22; Rockfinder/Getty Images, p.23; fotorobs/Shutterstock, p.24; Terry Moore/Stocktrek Images/Getty Images, p.25; Staff Sgt. Daniel Ter Haar/DVIDS, p.26; Stocktrek Images/Getty Images, p.27; Anadolu Agency/Contributor/Getty Images, p.28; U.S. Navy/DVIDS, p.29

Design element: SEAN GLADWELL/Getty Images; geengraphy/Getty Images

Cover: Ofer Zidon/Stocktrek Images/Getty Images